HOARD
L·E·G·A·C·Y
VOLUME TWO

HOARD'S DAIRYMAN

Published by W.D. Hoard & Sons Company
Publishers since 1870

Copyright © 2016 by *Hoard's Dairyman*
All rights reserved. No part of this book may be reproduced or transmitted in any form or by any means, electronic or mechanical, including photocopying, recording, by any information storage and retrieval system, or otherwise without permission in writing from W.D. Hoard & Sons Company, with the exception of brief excerpts in reviews or as provided by USA copyright law.

Editor — Maggie Seiler
Art Director — Ryan Ebert

Printed in the United States of America.

hoards.com

Library of Congress Control Number: 2016950162
Hoard's Dairyman
Hoard Leadership / Hoard's Dairyman
ISBN 978-0-932147-56-1

20 19 18 17 16   1 2 3 4 5
First Edition

W. A. Haard
Founder, 1885

Hoard's Dairyman editor, W.D. Knox, reviewing the magazine coming off the press.

Introduction

The history of *Hoard's Dairyman*, which first began in 1885 as a "journal devoted to dairy farming," has grown into a well-respected magazine and electronic communication entity with a legacy now spanning 13 decades. These days, English, Spanish, and Japanese versions of the magazine circulate in 80-plus countries around the world.

W.D. Hoard, the 16th Governor of Wisconsin, not only founded *Hoard's Dairyman*, but also transformed Wisconsin into America's Dairyland. Along the way, he started North America's first dairy school, led the historic tuberculosis eradication campaign, and began the first youth page for boys and girls on dairy farms . . . even before the creation of 4-H and FFA. Above all, Hoard was a friend to farmers of all ages with all breeds of dairy cattle.

Bill Hoard Jr., at the Hoard's Dairyman Farm.

Art director, Jim Baird, working on the magazine in the early 1990s

New Hoard's Dairyman Farm parlor built in 2007.

The *Hoard's Dairyman* staff continues to carry on Hoard's vision by leading a multitude of dairy causes. Over the years, these efforts have included leading a brucellosis eradication program, championing multiple component milk pricing, creating sketches to advance linear scoring in breed classification programs, and even authoring a book capturing the 50-year history of World Dairy Expo. Through the years, *Hoard's Dairyman* has remained committed to giving dairy farmers the most factual and current information available to assist them in making business and farm decisions. This is all based on recommendations rooted in science and practical experience.

2016 editorial staff

That is the very reason we created *Hoard Legacy* as a follow-up to *Hoard Leadership*. Our overarching desire was to inspire readers by selecting W.D. Hoard's most insightful quotes and pairing them with historic images to chronicle dairy innovation.

We hope you enjoy reading *Hoard Leadership* and *Hoard Legacy*, which we believe will soon become timeless dairy classics.

Corey A. Geiger
Managing Editor
Hoard's Dairyman

The fact is, this country needs a new race of farmers, men who look at the business from the standpoint of knowledge, training and skill. The smallest farm holds all of the problems; the largest holds no more.

1960 W.D. Knox elected president of National Dairy Shrine.

1960 National Mastitis Council is launched. *Hoard's Dairyman* editor W.D Knox asked to keynote formation meeting. First industry-wide effort to reduce dairying's most costly disease of dairy cattle.

"Real success is measured in broader terms than dollars and cents; it is based on an appreciation and proper use of the profit attained in bettering farm living conditions."

1961 W.D. Knox named by President of the United States to bipartisan National Agricultural Advisory Commission to advise White House on farm policy. Only editor in U. S. ever appointed.

1962 Jim Baird paints second Foster Mothers painting in a scene that features dehorned cows.

There can never be overproduction of fine butter and cheese. It is the poor stuff that costs just as much to make which clogs consumption and brings final loss.

Without a high standard, no advancement would be made in any endeavor.

1963 Golden Guernsey Co-op, Waukesha, Wis., is the first dairy in this country to pay producers on protein and fat.

1964 Christmas cover becomes an annual tradition on the December issue. Painted by art director James S. Baird, earlier efforts appeared on 1948, 1955, and then continuously since the 1964 issue. Baird's last Christmas cover appeared in 2007.

"Not only 'speak to a cow as you would to a lady,' but care for her as a true gentleman does for a lady; that is, if you want to make money at dairying."

"Tough as the year has been, corn fields fertilized by the cow show up a great growing crop."

1965 Class I Base Bill passes in Congress. Initiated by *Hoard's Dairyman*. It is the first amendment in 28 years to federal milk order legislation. Passage follows four years of struggle.

1967 Dairy Shrine names W.D. Knox Guest of Honor for his national leadership in the field of dairy cattle health, expanding the markets for dairy products, and development of peacetime dairy programs to improve the economic lot of the American dairy farmer.

"New beginners in the business of breeding registered cattle quite generally are looking around for something cheap. Let them continue in the work for a few years till they get their eyes open, providing they are real students of the art and science of breeding, and we will see them reverse the order of things. They will then cease to look for cheap animals, but rather the best animals."

1967 Controlled breeding is possible. This new research described a product and technique for synchronizing the estrous cycle of cattle.

1968 Publishes first ranking of A.I. sires on Predicted Difference for milk. It is a great leap forward in use of genetically superior sires, resulting in higher producing herds, and a more efficient industry.

The breeders of cows noted for great flow of milk can commend them to practical men, just in proportion as they show up a large percent of solids. Let them breed and feed for this end. The vein that simply shows a flood of low-grade milk won't pay to work as it once did.

1970 W.D. Knox named World Dairy Expo 1970 Industry Person of the Year. That same year, he receives Distinguished Service Award, from the American Dairy Science Association.

1971 Launched campaign proposing that milk be priced on protein content as a means of more equitable pricing. Would lead to even better quality beverage milk.

"The more I investigate this question of net profit, the more astounded do I become at the tremendous difference that exists between the men who think and those who will not think."

"The best cow in the world could not do good work unless well cared for and rightly fed."

1972 W.D. Hoard, Jr., passes away at age 75. Editor W.D. Knox becomes general manager and Eugene C. Meyer is named managing editor.

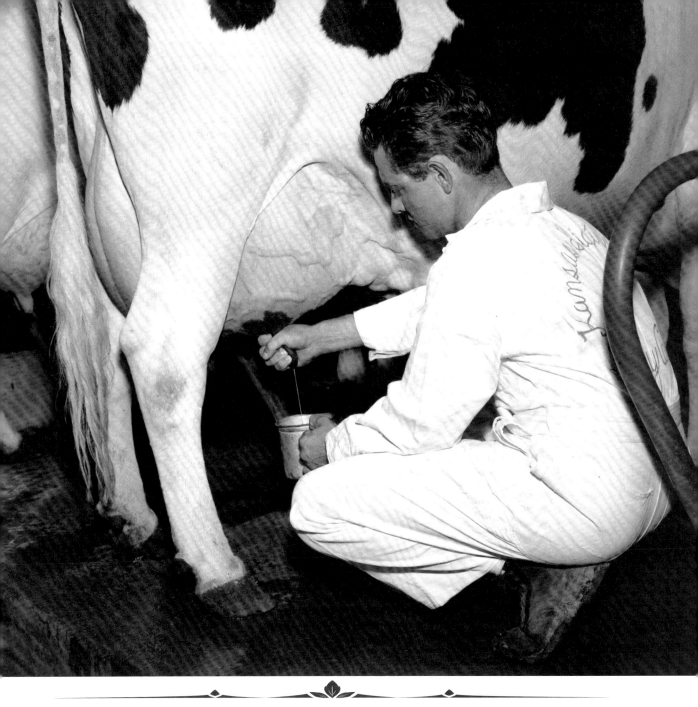

1972 Teat dipping and dry cow treatment helps eliminate mastitis: A Cornell study shows that dairymen could increase production 1,051 pounds per cow per year by using a mastitis control program.

"Every farm is a school, a laboratory, a court, where problems arise which demand study, understanding and good judgment in deciding. It is a small cheap man who thinks he is big enough to fully comprehend those problems without the aid of other men's knowledge. That is the reason why we see so many comparative failures in farming."

1973 Honorary Doctor of Laws degree awarded by the University of Wisconsin to W.D. Knox for his contributions to the welfare of one of Wisconsin's great industries, for his influential leadership in an important segment of professional journalism, and for his work on national agricultural policies.

1974 Editors urge the use of calf hutches to cut losses due to inadequate ventilation and sanitation. Put into use at Hoard's Dairyman Farm. Now used to raise calves on one out of every three farms in the U. S.

" Now, how shall you reduce the cost of production? By increasing milk production per cow and keeping fewer cows."

" Women are taking hold of dairying, and we are greatly impressed with their success in mastering its difficulties. Women are more patient of small details, and dairying is full of such details."

1974 First embryo transplants eligible for registration: On March 12, 1974, a grade Jersey gave birth to a Holstein bull calf, the first Holstein to be born as a result of an embryo transplant eligible for registration with the Holstein-Friesian Association of America. (shown right)

1975 U.S. milk production per cow averaged more than 10,000 pounds for the first time (10,360). DHI cows averaged 13,632.

"Hoard's Dairyman does not deal with politics except when politics threaten the integrity of dairying."

"The superior cowman is one who makes a close study of each individual cow in his herd and shows his ability by adapting himself and his treatment to the individual needs of each cow."

1976 President of the United States names W.D. Knox to Advisory Committee on Trade Negotiations. Sole representative of dairy industry. Only editor in U.S. appointed. Knox served under Presidents Ford, Carter, and Reagan. Prior to that, Knox served on the National Agriculture Advisory Commission.

1977 New Zealand government asks W.D. Knox to be its guest in study of its dairy industry. Reports on New Zealand and Australian role in world trade in dairy products, impact on U.S. dairymen.

Now all there is to this thing we call progress in agriculture, the science of the thing, is a study and knowledge of principles, the law that governs soil life, plant life, and animal life. What a tremendous advantage a man's hands have in the struggle when he has a well-stored mind to guide them in their work.

1978 U.S. Secretary of the Interior names Hoard's Dairyman Farm to "National Register of Historic Places." (shown right)

1980 Managing Editor E.C. Meyer receives Distinguished Service Award, from the American Dairy Science Association.

"As farmers, we must constantly keep in mind that there is no animal on the farm that possesses the power to transform course forage and grain into the highest order of human food like the cow. But the lesson for us to consider is how we can best aid this wonderful animal in her transforming power."

1981 — Dairy Shrine dedicated in Fort Atkinson, Wis., home of *Hoard's Dairyman*. National dairy leaders selected site for long-desired museum and historical exhibit. Managing Editor E.C. Meyer led final drives for funds, making the Dairy Shrine a reality.

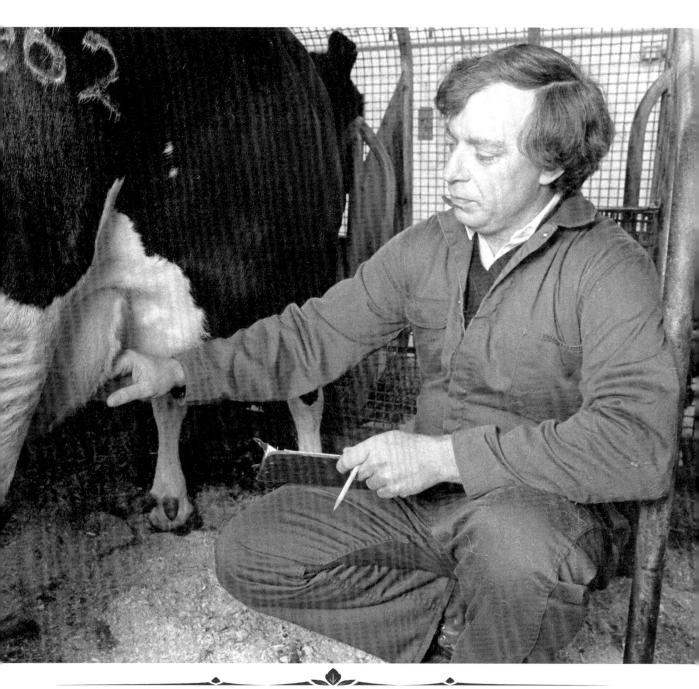

1982 Publishes first interview with Cornell's Dale Bauman, whose research on bovine growth hormone reveals a 10 to 40 percent boost in production when administered to cows fresh 60 to 100 days.

Bacteria are small affairs but if of the wrong kind can make a lot of trouble.

The dairyman is in commercial relations with the cow in two places, first, as to the character of the cow, and second, as to the character of her feed.

1982 Cheese yield pricing plan advances as an incentive to produce milk with high protein content rather than discouraging high solids, as do most payment plans.

1983 Veterinary Editor Dr. Lee A. Allenstein is honored as the U.S. Bovine Practitioner of the Year.

There are multitudes of cow keepers, but how few proportionately are dairymen in fact. The distinction between a cow keeper and a dairyman is as broad as that between a woodsawer and a carpenter.

1983 Magazine recognizes Johne's disease as a national herd health issue and enrolls the Hoard's Dairyman Farm in a pioneering control program.

1983 Is methane from manure for you? This was our first look (January 10, 1983) at what biogas production involves, what size system is needed, how the gas can be used, and its value for fuel or electricity.

For hundreds of years the principal purpose of schools and their resultant education was to make professional men. It was not thought that the farmer needed any special schooling and mental training. But a change of opinion and judgment has come.

1983 *Hoard's Dairyman* survey shows strong support for Dairy Surplus Reduction Act of 1983.

1984 *Hoard's Dairyman* gives strong support to proposal for dairy promotion order that would provide a 15-cent-per-hundred weight checkoff for dairy promotion and research.

> Here are the three cardinal principles of dairy farming, good land, good crops, good cows, but at the bottom lies good land and a farmer wise enough to keep it good.

> Your fortune and mine does not depend upon the gross amount of money that goes through our hands, but upon the percent of profit.

1986 Publishes the first ranking of dairy cooperatives. Annual list sheds light on cooperative's share of national milk marketing.

1986 Gary Vorpahl succeeds Mark "Bud" Kerschensteiner as head of marketing and advertising for *Hoard's Dairyman* after 41 years of service to the company. Kerschensteiner is the last direct relative of Governor Hoard to work for the company.

"We know of several ladies in the United States and Canada who are successful managers of dairy farms and also breeding herds of dairy cattle. Indeed it is very rare to find a woman who is not successful when she sets out to manage a farm or any other kind of business."

1986 Editor E.H. Row is named president of prestigious National Mastitis Council.

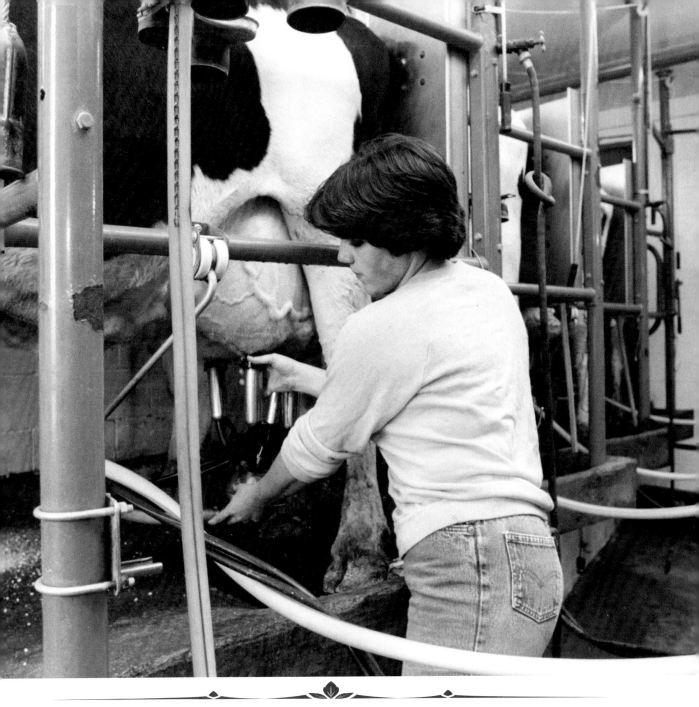

1986 Somatic cell count limit for both Grade A and manufacturing grade milk is reduced from 1.5 to 1 million on July 1, 1986.

"Dairy farming is, of necessity, a business of brains, a business not of rude strength so much as patient skill. A shrewd, intelligent dairy farmer will take pains to surround himself with intelligent help. He will be willing to pay well for the right sort of a man for the wrong man is wasting his substance every day.

1987 Editorial calls for merging of United Dairy Industry Association and National Dairy Board.

1987 Following years of urging through editorials, USDA includes protein pricing in a recommended merger of the Great Basin and Lake Mead federal milk marketing order.

"It requires keen perceptions; watchful care; a kind, humane heart; and abundant energy to succeed with a herd of cows."

"Of course, this is no excuse for feeding rotten or moldy silage, or rotten or moldy feed of any kind."

1987 Predipping gains attention as a mastitis and somatic cell count reducer.

1988 Managing Editor E.C. Meyer is honored by World Dairy Expo as Industry Person of the Year. Just two years earlier, he was named National Dairy Shrine Guest of Honor.

"I want to get at the man that produces the milk. I want to see that man's profits enlarged. I want to see his labors lightened. I want to see his intelligence increased. I want to see his family happier and his home more cheerful, and the man, and all that belongs to him, a better product of this day and civilization. That is what I want to see."

1988 Multiple component pricing becomes a reality in the Great Basin federal order, on April 1, 1988.

1988 Michigan dairymen are taking the siding off of their freestall barns and replacing it with curtains.

"There is no vocation on God's green earth that calls for higher elements of character, for deeper research, for grander nobility of nature than that of the farmer."

"The prosperity of the city is bound up in the prosperity of the farm."

1989 "Cover blurbs" are added to tell readers what is available inside.

1989 *Hoard's Dairyman* launches its Japanese edition.

"Since the inception of Hoard's Dairyman, it has been our policy to advocate practices and principles which we think will be to the dairyman's interest. We do not dodge issues, but meet them squarely, and our columns are open, to a reasonable extent, to anyone who desires to take issue with or support us."

1989 Ewing H. Row succeeds Eugene C. Meyer as managing editor, member of the editorial staff 41 years.

1989 Editors campaign against state or federal bans of controversial bovine somatotropin, pending approval by Food and Drug Administration. Unpopular position angers many readers, and some of them cancel their subscription.

The really practical breeder is, as well, a developer of his animals along chosen lines. In breeding dairy cattle, the true lines are milk production and health and constitution. We can breed for health and constitution as well as breed away from it.

1990 Editorial calls for labeling A.I. sires by type of progeny test. Leads to new coding system adopted by National Association of Animal Breeders.

1991 Baird paints third edition of the Foster Mothers collection and adds the Milking Shorthorn.

Remember 'knowledge is power' for
the dairy farmer as well as everybody else.
Don't be afraid of knowing more.

I wouldn't give much for anybody in the routine
of life that had no heart in the life they live.

1991 W.D. Knox is requested to lead off testimony on national dairy policy in U.S. Senate hearings.

1992 Jim Baird, magazine's art director, retires after 44 years. Widely known for cover photographs, Foster Mother paintings, and Christmas covers.

Make a thorough study of your work. See what it means to the man who pays you for it and particularly what it means to the livestock you are ministering to every day. Cultivate a love for your work. If you hate your work, you are cutting the ground from under your own feet.

1992 Magazine encourages and supports reporting of commercial dairy farms – those holding licenses to sell milk. USDA takes over reports in 2005.

1992 U.S. dairy cows produce 0.2 percent of the world's methane, one of the gases blamed for global warming. Even termites produce more (0.5 percent), says J.B. Holter at the University of New Hampshire.

Preservation of soil fertility must be the first object of all truly profitable farming. Let every reader of Hoard's Dairyman look over his farm methods carefully and measure them by that question.

There is no finer element in American society than the intelligent, well-cultured farmer.

1993 Animal Industry Foundation warns that every organization and dairy operation should be prepared to respond to anti-animal activists.

1993 The "Got Milk?" slogan is created by Goodby Silverstein & Partners for the California Milk Processor Board.

'Teach me, O! Lord the value of patience.' Well, it is worth just as much in dairy farming as in religion. Patience is the key that unlocks the secret of the soil, the undeveloped motherhood of the cow, and the earnest loyalty of the men who labor for you.

1993 Food and Drug Administration approves bovine somatotropin.

1994 Baird paints fourth Foster Mothers of the Human Race.

"You can't do much in any enterprise without the right man at the helm. Even a cooperative corporation must have the right man or it goes to pieces.

"If every dairyman would keep a glass and inspect the excrement from his animals (looking for undigested grain), he would have an opportunity to learn a very profitable lesson.

1994 The National Dairy CHeckoff came under one umbrella by the creation of Dairy Management Inc., (DMI). DMI started out as a joint venture between the National Dairy Promotion and Research Board and the United Dairy Industry Association.

1994 *Hoard's Dairyman en español* is launched. Abelardo Martinez, D.V.M., named first editor.

"Cows not only vary in the amount of milk and its fat content but also in the quantity and value of the feed consumed to produce milk. While it is true that small yields are seldom profitable, it is also true that the largest yields do not always return the most profit."

1995 Promotes the collection and reporting of actual producer take-home pay for a number of years. In 1995, USDA begins reporting mailbox prices.

1995 Wisconsin researchers develops a procedure for synchronizing the time of ovulation using gonadotropin-releasing hormone (GnRH) and prostaglandin. Program later referred to as ovsynch.

"I have given years of study to the dairy cow and I believe a good deal about her, but more and more I am convinced that the darkest place in the world is the inside of a dairy cow. Chemists have their laboratory. Mechanics may have their machines, but no man knows how the dairy cow transforms the hay and grain she eats into milk."

1996 Magazine launches website, www.hoards.com. Provides readers the latest market information and breaking news.

1997 Number of dairy farms with a license or permit to sell milk falls below 100,000 for the first time.

A 40-acre farm that can feed one grown animal per acre is a bigger farm than one 160 acres in area that can feed no more – and the man who operates is ever so much the bigger man, measured by what he really does.

Underconsumption is a thousand times more of a calamity to mankind than overproduction.

1997 W.D. Knox is named to inaugural group of American Dairy Science Association Fellows.

1998 S.A. Larson (seated) succeeds E.H. Row (standing) as managing editor. Row served readers of *Hoard's Dairyman* for 31 years.

"No other animal on the farm is drawn upon so greatly and gets so little for what she has to do as the cow. We demand of her that she give milk enough to make a profit; that she produce a healthy, vigorous calf and that she support herself in good health. No wonder the Hindoos considered her as one of Heaven's most precious gifts to men."

1999 D.J. Halladay is named first editor of *Hoard's WEST*, a publication specifically written for Western readers.

1999 The first commercial robotic milkers installed in North America.

Farm life has splendid possibilities for the most perfect life, the most invigorating, the freest, and most inspiring. Its work is not drudgery when the significance of that work is comprehended.

2000 Editorials lead to national review and revision of dairy export-import policy.

2001 *Hoard's Dairyman* and longtime contributer Ev Thomas partner to publish the first known article cautioning about the excessive use of copper sulfate in footbaths.

Some day we hope farmers in general will conclude that farming is as much a work of brains as law, medicine or any of the so-called learned professions. But everywhere in all callings, there are men who follow the 'line of least resistance.'

2002 Editorials and articles on showring ethics bring issue into limelight . . . results in stepped-up enforcement and youth education efforts. (shown right)

2003 Magazine throws support behind Cooperatives Working Together . . . an industry-wide, self-help program to improve dairy farm incomes. Magazine becomes primary sponsor of National Dairy Quality Award program.

If we are to be the men we should be, and especially if we are to make good farmers of ourselves, we must enjoy our work. It must be welcome to our hands and hearts. It is a duty every man owes to himself to train his mind to see the good side of everything he comes in contact with.

2003 Sexed semen becomes commercially available. The sperm-sorting technology results in 90 percent heifer calves. (shown right)

2004 By working with Cornell Extension's Tom Kilcer, the magazine reports that wide swath hay makes more milk. Also wide swath hay leads to quick dry-down, same day harvest.

Not how to consume less but how to produce more on less land is the present and continuing problem for the farmer.

Turn your attention towards reducing the cost of production so that there is a margin of profit, even at low prices.

2005 Brian Knox succeeds his father, W.D Knox, as president of W.D. Hoard and Sons Company, the fifth person to lead the company since 1870.

2005 — Retired Managing Editor Eugene C. Meyer and editor/publisher W.D. Knox pass away within three months of each other. Combined, they served Hoard's Dairyman and the dairy industry for 121 years.

The milk market will never be in a satisfactory condition until buyers recognize that all milk is not alike and become willing to pay better prices for better qualities. The fact that practically all milk today sells for the same price is the chief force in reducing the average quality of milk in the state.

2007 Managing Editor Steve Larson is named World Dairy Expo 2007 Industry Person of the Year.

2007 *Hoard's Dairyman* recommits itself to the dairy industry by constructing new state-of-the-art dairy facilities to help serve its readers more effectively.

In this controversy between the show ring and the milk pail, there is one thing that impresses us and that is that nothing keeps up the selling value of land and cows like 'producing ability.'

2008 *HD Notebook*, an online dairy blog begins providing news to readers at www.hoards.com.

2008 *Hoard's Dairyman* publishes USDA's research on genomic predictions which predicts, how bulls and heifers will perform based on DNA tests.

This tide of progression, ought to convince us all that dairy farming is a business of the brain, that just in proportion as the farmer keeps his mind alive and quick to comprehend dairy truths will he grow and prosper in the work.

2009 Today's dairy system produces considerably less manure (24 percent), methane (43 percent), and nitrous oxide (56 percent) per billion pounds of milk than the 1944 system. The current carbon footprint per gallon of milk is one-third as great as it was 65 years ago.

2009 Hoard's Dairyman Farm adds Jersey cows to the existing Guernsey herd.

"Fortune does not knock at the door of the cow man, and wait for him to come out and welcome her. He must make his own tide of fortune, and do it by hard, intelligent work."

"Corn silage and good alfalfa hay about solve the problem of a cheap ration for cows. It needs but a small ration of grain to keep them in full flow of milk."

2009 A new study, funded by the National Dairy Council and published in the *Journal of Nutrition and Metabolism*, helps connect dairy consumption to weight maintenance following weight loss for overweight and obese adults.

2010 Japanese Editor Tadanaga Komori (center), of the Japan Livestock Trading Corporation, is named World Dairy Expo International Person of the Year. Standing to the left of Mr. Komori is Spanish Editor Dr. Abelardo Martinez, and on the right is American Editor Corey Geiger.

"If you have a good cow and she will pay for what feed she takes, sell it to her. Sell her all she will pay for. Then investigate, not guess. Keep a record to know whether she is paying or not."

2010 State of Wisconsin creates state holiday designating October 10 as William D. Hoard Day.

2010 *Hoard's Dairyman* celebrates 125th anniversary. Redesigns magazine and website. Commissions Minnesota artist Bonnie Mohr to paint new "Foster Mothers of the Human Race."

The objective of dairy study, dairy reading, dairy thought, and dairy practice is to look into this question of a principle. When we once get it clearly into our comprehension, the principle then guides us and not blind practice.

2011 Steve Larson is named National Dairy Shrine Guest of Honor in recognition of editorial leadership and high regard in the industry as a counselor for a wide range of dairy organizations. (shown right)

2012 Spanish editor and publisher of *Hoard's Dairyman en español*, Dr. Abelardo Martinez, Tlalnepantla, Mexico, is named World Dairy Expo International Person of the Year.

"A dairy cow must have care. She is sensitive, she is nervous, she has a peculiar susceptibility to change, to cold, to warmth, and you ought to write over every cow's stanchion the words "comfort" and "care" in large letters. Your pocket book is at stake."

2013 Introduces *Hoard's Dairyman Intel*, a weekly electronic newsletter with uniquely written content to supplement the print magazine's coverage.

2013 Corey A. Geiger succeeds Steven A. Larson as managing editor. Larson was a member of the editorial staff for 44 years and now serves as editorial consultant.

"Wherever we look we find that the good farmer of today is the man who is most wisely farming for the future. Upon his mental ability to foresee and provide for the future will rest the largest reward for his labor in the present."

2014 Investigative reporting on rural domestic violence by People Side columnist Andrea Stoltzfus is read into the Congressional Record by Senator Patrick Leahy (D-VT).

2015 The parent company of *Hoard's Dairyman*, W.D. Hoard & Sons, acquires *Hay & Forage Grower* magazine. Mike Rankin is named managing editor.